Study Essays on the Advance Sciences

Miguel A. Sanchez-Rey

Table of Contents

Gauge Theory is a Brain Drain
5-10

Review of, *The Workshop and the World*
12-18

Freedom of Speech or Curtailing the Freedom of Speech?
20-25

The Leading Professorship and the Fellowship
27-36

Gauge Theory is a Brain Drain

The Leading Professor Miguel Angel Sanchez-Rey [*The Grandmaster, The Master of Space-Time*]
The Academy of Advance Science and the Technological Sciences

The history of mathematics is fraught with mathematical Platonism that seems tangible and beautiful but is bound to blow up in the face of a mathematical genius. Gauge theory -- the mathematical theory that the Lagrangian is invariant under particular Lie groups of local transformations, underpins much of modern physics in the language of differential geometry.

From classical modern physics (gauge transformations) to contemporary quantum field theory (connections), gauge theory has allowed theoretical physicists and mathematicians to delve into the very nature of space-time, particle interactions, and soap bubbles, i.e., the gauge symmetry of James Clerk Maxwell's field equations, the $SU(3) \times SU(2) \times U(1)$ Standard Model of the electroweak, nuclear, and strong forces, and/or the quantum thermodynamics of nuclear plasma.

Permitting substantial progress in geometry, i.e., geometric analyses, Chern-Simons form for the Jones Polynomial, and/or perturbation theory, gauge theory has proved to be an important and yet overuse sub-branch of mathematical engineering.

Whether through path integrals and/or non-linear partial differential equations, the atypical and redundant is misconstrued as functional sophistication and procedural eloquence in computational decision-making.

Poor in its construction and counter-intuitive in its very application to the compactification of higher dimensions, i.e., Kaluza-Klein reduction of gravitation and electromagnetism, it's an atypical redundancy that drained the field of physics of needed computational power at the expense of maximum efficiency. True, non-abelian gauge theory proved to be worthy of the peace prize in mathematical physics. But gauge theory means different things to many other theorists.

Overall, it means a brain drain that led to the blow up of much of mathematical geometry: more its emphases, the more its overuse -- more its overuse, the likelier that the field of modern physics will succumb to mathematical incapability and experimental inadequacy.

Eventually soap bubbles pop, yet too much of a good thing can be a bad Platonic idea. It's more than just over for the geometric principle, it's complete overkill for the field of mathematical theory.

The parsimony, elegance, and simplicity of Advance Physics.

Review of, *The Workshop and the World*

The Leading Professor Miguel Angel Sanchez-Rey [*The Grandmaster, The Master of Space-Time*]
The Academy of Advance Science and the Technological Sciences

The scientific process has no infrastructure. Yet the shapers of said infrastructure are the arbiters of authority in the scientific process. Authority in the sciences is a peculiarity of the sciences. The only authoritative response is that authority is a weighty conundrum. Yes, scientists count themselves as authorities in many different academic fields: physics, mathematics, philosophy, and/or etc. Presenting themselves as the experts of their chosen field. With qualities of leadership and a revered persona in which others will look up to their stature to seek their input and expertise.

Though the authority figure is conjectured as a necessary academic qualification, academics are in no part, de facto, authorities in the scientific process. Rather, though the scientific infrastructure is modeled in the experimental and peer-review process, with accolades and higher degrees that are awarded base on merit and achievement, the remaining academic hierarchy has dictated that the highest authority is no longer a practical truism.

There is no telling that the planet has seen the collapse of the academic hierarchal order: plunging the scientific process into disarray. Yet being a scientist does not mean that appealing to authority is a tautology, rather the authoritative figure is bound to be a misleading and detestable figure-head whom may overstep their academic reach in the sciences.

Such a figure-head may resort to academic dishonesty to pose honest research and results. Yet be known as a cult-figure with a large following that utilizes the information age to give the persona of fame and expertise in a global arena. There is no doubt that the workshop is a laboratory of scientific impotence: with painstaking end results that are observed to be even questionable to a peasant.

One agrees, omnipotent authority in the sciences has met its solemn demise. But being the authority does not mean he and/or she is the shaper and maker of a questionable infrastructure.

The only indication is that the highest authority is a model of the sciences that is near emblematic of higher qualification. Such higher qualification dictates that the model of the sciences is an expert decision-maker that far surpasses the authoritative figure-head, but whereby higher learning means the capability to make inform and educated decision-making that leads to strategic and tactical genius in their chosen field.

They are exemplary models but hold no complete jurisdiction in their chosen fields, rather the jurisdiction lies in the scientific process of the winning strategy (in the theoretical, experimental, and applied sciences). A powerplay of international collaboration that seeks to impart independence and confidence to achieve near-perfect decision-making in their chosen field of higher learning. To become world leaders and renowned experts in the sciences -- not logical misconceptions.

Which the workshop is no more indicative of, instead the workshop becomes a sanctuary for scientific exploitation and fallacious logic.

Freedom of Speech or Curtailing the Freedom of Speech?

The Leading Professor Miguel Angel Sanchez-Rey [*The Grandmaster, The Master of Space-Time*]
The Academy of Advance Science and the Technological Sciences

The executive order has been signed that asserts the freedom of speech on U.S. campuses. It's deplorable that not enough has been said about what the freedom of speech implies for a dystopic reality. It seems credible that academics are to assert their right to free speech but should all those involve in higher education, i.e., faculty, student, staff, etc., assert their freedom of speech to gain and apply knowledge about the Scientific Age?

Granting access to higher education requires that students meet academic standards and when those standards aren't satisfied, said students are dismiss from academic studies. Professors are designed to impart specialized knowledge in the form of gathering amongst students. To make certain that students gain competence in their chosen field until they reach confirmation of their undergraduate degree (A.B., B.A., B.S.). After which they will choose to pursue a graduate degree (M.A., M.B.A., Ph.D., M.D.), research position, and/or employment in the private and/or government sector (or other formulation of preoccupation). While professors instruct class, further their specialization and/or gain the opportunity to publish through peer-review and/or a publishing house (for humane purposes).

Such universities and/or colleges are defined as waypoints. Launching grounds for the elite colleges -- in which, their alumnae will set out to take part in cutting-edge research and elite decision-making (at a national or at a global scale).

Freedom of speech is not an inalienable right in the Scientific Age, instead only so few are ever more than qualified to discover, examine and/or construct wild ideas. And since wild ideas can only be ascertain and establish by an expert in his academic field -- at a certain point, others will learn to accept those ideas (once the Scientific Age has determined their relevance to the scientific process).

Not everyone can tolerate wild ideas. And the risk that poses in asserting the freedom of speech -- at the university level, is that not all students can pass the class curriculum. Even then, not all can embark on the scientific process.

The Scientific Age is an uncontrollable scientific machine that is to be tempered, but not so easily overburdened by benign ideas that may eventually cause havoc to the sciences. Imposing a lockdown on wild ideas will mean that much about the contemporary era is to be forgotten after the Scientific Age. But there is no escaping futility: as all things eventually come to an end.

The Leading Professorship and the Fellowship

The Leading Professor Miguel Angel Sanchez-Rey [*The Grandmaster, The Master of Space-Time*]
The Academy of Advance Science and the Technological Sciences

The fellowship is an association of scientists, in the United Kingdom (U.K.) and the Commonwealth, that seek to communicate progress in the natural sciences (implemented by King Charles II in the Royal Society Charter). Encouraging traditional meritocracy, the fellowship is symbolic of academic prestige. Yet as a premier learned society, the fellowship aims to assert themselves as the gatekeepers of the natural sciences: validating scientific results and proclaiming scientific achievement.

By nature, public intellectuals: the fellowship of the Royal Society has been relatively moderate in their viewpoints about the natural sciences. Yet radical science has exposed the fellowships culpability to political radicalism, i.e., utilizing extreme tactics that furthers neo-Fascist policies that are in opposition to the Magna Carta.

For those reasons, the fellowship poses external risks to the leading professorship. While the fellowship is motivated by class consciousness, the leading professorship mandate is to commit righteous acts. Though British affairs is the paradigm of the fellowship (located not only in the United Kingdom but also scattered throughout the Commonwealth of Nations. Where foreign fellows of the Royal Society reside in their home countries outside the U.K. and the Commonwealth), the leading professorship oversees *The Academy of Advance Science and the Technological Sciences*: a global scientific academy that aims to impart leadership position -- by means of higher learning, to achieve strategic ends.

Yet as the fellowship is devoted to a monarchal traditional system, the leading professorship is incline to political federalism. Whereas the fellowship is of competence to the natural sciences, the leading professorship is of higher qualification to the sciences. Even then the fellowship is irrelevant to long-term global military interests. The fellowships contemporary motives are to promote the British system and to protect neo-capitalist meritocracy, for purposes that are non-democratic.

Inept of authoritative action, the fellowships strength lies in its capacity for scientific advocacy. While the leading professorship is to carry out its giving mandate to reach a logical conclusion that is consistent with scientific procedure. Modeling itself as a powerhouse, the fellowship lacks the discipline and the impetus to motivate the scientific process to its schematic consequence.

The fellowship is of little concern to the leading professorship. Boundaries are to be enforce and indirect means is to be implemented to secure its mandate. Stressing such mandate, the fellowship is both incapable of long-term strategic planning and a liability to the ranking system.

But the fellowship strength lies in advocating for the sciences. Yet the Royal Society must find relevance in the Scientific Age, while the factual nature is that the fellowship is a product of the British empire. But disregarding their controversial self-image, the capacity for the fellowship to commence as, "chess masters" can become their greatest strength. To educate and prepare new generations of citizen scientists.

Education and preparation that will set into motion the acquisition of expertise in decision-making: coinciding with both early achievement in formal systems, the sciences, and excellence in civic-participation. Yet more accomplished than a typical game theorist.

Eluding to a heavier burden for the fellowship than any other learn academy.

The Leading Professorship is design for top-military scientists that've completed the PHPR Protocol (further stipulated by both The Guiding Principle and Procedural 6th). Yet the fellowship is entitled to recommend candidacy after the utmost requirements are satisfied. Once the leading scholar is selected, he and/or she is granted the professorship until such stipulation (or the utmost requirements) is violated.

When such stipulation (or the utmost requirements) is violated, the fellowship is denied any further selection of candidacy until disciplinary action is completed by their prospective war-college, i.e., pending a strenuous criminal investigation by the intelligence agencies (under the jurisdiction of the North Atlantic Treaty Organization [NATO] strategic command).

Upon election to the Royal Society, the leading scholar is to cease direct diplomatic channels with *The Academy of Advance Science and the Technological Sciences* (further implied by stipulation).

Bestowing reason, clarity, and civility to its mandate by bringing academic prestige to the Royal Society -- by way, of its advocacy of the scientific process. Whereby the scientific process is indebted to an act of higher qualification.

Begin Communique

To the Nobel Prize Peace Committee at Oslo, Norway:

You do not dare violently torment, suppress, and/or murder a saint…

For I am not a saint…

And yet I am not a sir.

I have been dishonored by her majesty the queen. But I did an honorable thing: for my own sake and for the sake of humanity…

The Nobel Prizes have been compromised by ulterior elements that pose an existential threat to the international order…

PHPR is off-limits to the prize system in the long-term. The AASTS is not open for public discussion.

Obey the international order…

Obey NATO law…

Otherwise liable to charges of war-crime, espionage, and/or high-treason…

Under the penalty of death by hanging and/or death by firing squad. Without hope of lethal injection…

I am one of many things but first and foremost:

The Leading Professor Miguel Angel Sanchez-Rey [*The Grandmaster, The Master of Space-Time*].

PHPR is top-secret and the AASTS is now classified…

It's the beginning of Starfleet. As close to a united federation…

By protocol, directive, and by mandate…

The Scientific Age is an age of wild anticipation…

Make no further attempts to communicate and/or incite hostility…

End Communique

www.ingramcontent.com/pod-product-compliance
Lightning Source LLC
Chambersburg PA
CBHW081146170526
45158CB00009BA/2715